ASPECTS OF G

General Editors: J. H. Johnson ...

Weathering
and Landforms

CLIFF OLLIER

University of New England, New South Wales

SECOND EDITION

Nelson

Thomas Nelson and Sons Ltd
Nelson House   Mayfield Road
Walton-on-Thames   Surrey
KT12 5PL   UK

51 York Place
Edinburgh
EH1 3JD   UK

Thomas Nelson (Hong Kong) Ltd
Toppan Building 10/F
22A Westlands Road
Quarry Bay   Hong Kong

Thomas Nelson Australia
102 Dodds Street
South Melbourne
Victoria 3205   Australia

Nelson Canada
1120 Birchmount Road
Scarborough   Ontario
M1K 5G4   Canada

© C D Ollier 1974, 1990

First published by Macmillan Education Ltd 1974
Second edition 1990
(under ISBN 0-333-52067-X)
This edition published by Thomas Nelson and Sons Ltd 1991
ISBN 0-17-448186-1
NPN 9 8 7 6 5 4 3 2

All rights reserved. No paragraph of this publication may be reproduced, copied or transmitted save with written permission or in accordance with the provisions of the Copyright, Design and Patents Act 1988, or under the terms of any licence permitting limited copying issued by the Copyright Licensing Agency, 90 Tottenham Court Road, London W1P 9HE.

Any person who does any unauthorised act in relation to this publication may be liable to criminal prosecution and civil claims for damages.

Printed in Hong Kong.

Acknowledgements

The author and publisher would like to thank the following for permission to use photographs in this book: I. do Amaral, Figure 12; J. Field, Figure 29; D.K. Holdsworth, Figure 7; N. King Huber of the US Geological Survey, Figure 14; J.N. Jennings, Figure 4; G.F. Matthias, Figure 36; New South Wales Information Service, Figure 26; J.A. Peterson, Figures 22 and 25.

Cover photo by kind permission of N. King Huber of the US Geological Survey.

Contents

Preface	iv
1. Introduction	1
2. Weathering of basalt	3
3. Weathering of limestone	6
4. Weathering of granite	10
5. Weathering of sandstone	16
6. Holes, hollows and honeycombs	17
7. Weathering at the coast	21
8. Weathering in cold climates	23
9. Weathering in deserts	26
10. Weathering in the humid tropics	30
11. Weathering and life	32
12. Weathering, soil and slopes	35
13. Weathering of towns and tombs	43
14. Weathering products of commercial value	46
15. Weathering and landscape evolution	49
Topics for discussion	52
Project work	53
Bibliography and further reading	55

Preface

In recent years, geography has been changing with great speed. It is not primarily that the basic facts of geographical distributions are themselves changing, although, of course, this has happened. It is far more that geographers have come to think differently about the significance of geographical distributions, about how to study them, and about what topics are worthy objects of geographical investigation.

Nobody can remain in close touch with the expanding frontier of geographical knowledge at all its points. New developments in geography are the subject of contributions to learned journals, but these are difficult to track down and, even when found, they are difficult for the non-specialist to assess. Nor can all new developments be taken up quickly by the standard textbooks, which must necessarily go some years between revisions. As a result, *Aspects of Geography* has been organised as a series of concise reports by writers who are in contact with a particular sector of the subject's development. Although the series is particularly aimed at A level students and their teachers it is hoped that the series will also be useful to college and university students as an introduction to the various specialist fields that will be covered.

It is now several years since an examination question described weathering as the 'Cinderella of geomorphology'. Nevertheless it remains a neglected topic, all too readily seen as a trivial addition to the geologists' story, or a difficult, and intrinsically dull, prologue to geomorphology. Yet weathering is a critical part of landscape development, and most of the rocks the geomorphologist is concerned with are modified by weathering long before they are attacked by erosional processes. Further, while most of the control of denudation by weathering is indirect − through the thickness and nature of the regolith affected by erosional processes − on steeper slopes the control is direct since the limiting process is the rate of weathering, not the rate of erosion.

Cliff Ollier begins his book by noting the relationship between landforms and weathering with examples from tors and tombstones. These are of great interest and tell us a good deal about how weathering works. But as he reminds us at the end, it is the widespread preparation of rock for erosion that is the prime geomorphological significance of weathering.

PREFACE

This second edition has been extensively revised. The author has expanded and updated the work and new illustrations have been added. As well as revising existing chapters, new material has been included on weathering of sandstone; holes, hollows and honeycombs; weathering in humid topics; and weathering, soil and slopes. The author has also provided the student with ideas for project work which schools will find extremely useful.

J.H. JOHNSON
IAN DOUGLAS

1. Introduction

A marble statue, a house, or the paint on a new car will all become corroded and changed if exposed to the weather, a process so obvious that 'weathering' is a term used in everyday language. Rocks also suffer from weathering, and the breakdown of rocks in this way produces many distinctive and interesting landforms.

Many rocks are formed deep in the earth, under conditions of high temperature, high pressure, and the absence of large quantities of water or air; they react at the earth's surface to come into equilibrium with the conditions of low temperature, low pressure and the presence of air and water. The original rock is altered by weathering processes into weathering products, and the change may be accompanied by the formation of weathering landforms. 'Weathering' in this sense is extended to the alteration of rocks not only exposed to 'weather' but to alteration at considerable depths.

Weathering processes can be thought of in two categories: physical weathering breaks rocks down into smaller fragments, and chemical weathering alters the minerals in the rocks into different materials. Most rocks are made up of silicate minerals which are compounds of silicon, oxygen and other elements. These are commonly altered to clay minerals as weathering proceeds. Note that the word 'clay' is used in two different senses: first, to mean a material composed of particles smaller than 0.002 mm; secondly to mean a distinct group of minerals, the clay minerals, such as kaolinite, montmorillonite and illite. Some weathering products are carried away in solution and some minerals such as quartz may resist alteration completely.

Some rock is extensively altered chemically without any change in volume (isovolumetric weathering), and is known as saprolite. Rock structures such as joints, faults and quartz veins are frequently present to show that the rock has not changed volume despite extensive mineral alteration. Other weathering products, such as the insoluble residue from limestone solution, may remain more or less in place but without preserving rock structure. Some weathering products may be transported by hillside creep or wash, or mixed up by animal burrowing or root growth. Quite often it is difficult to decide whether a particular deposit is in place or not, and the word regolith is used to connote the entire surficial deposit of weathering products, regardless of whether they have been disturbed or transported.

INTRODUCTION

A vertical section through the weathered material, from ground surface to bedrock, is known as the weathering profile. It may be divided into several zones. The upper part is a leached zone, and there may be a precipitation zone below. The upper part will usually be oxidised, and there may be a waterlogged, reduced zone below. Other properties such as colour, mineralogy, or corestone abundance may be used to distinguish zones of weathering. In some places the regolith is very deep: depths of tens of metres are common, and hundred of metres have been reported from many parts of the world.

To illustrate these concepts we will first look at the weathering of several common rocks.

2. Weathering of basalt

Let us start by examining the weathering of basalt, a dark rock that is produced by the cooling of a lava-flow. It is a fine-grained rock, which means that the mineral grains that make up the rock may be too small to be seen with the naked eye. Many lava-flows have been observed in historic times, and by studying such flows we can get an idea of how much time is required for weathering to take place. Lava-flows are usually cracked in various ways, and the cracks or joints provide lines of entry for water and air, and hence weathering.

Basalt consists largely of two kinds of silicate mineral: feldspar, a silicate of aluminium, sodium and calcium; and augite, a silicate of calcium, iron and magnesium. Primary silicate minerals weather to clay minerals, with the release of some components to solution and the simultaneous formation of hydroxides of iron and sometimes of aluminium.

Clay minerals are distinctive minerals consisting essentially of layers of oxygen and aluminium (aluminate layers) together with water and metallic atoms called cations or bases. Some clay minerals such as montmorillonite have a structure of two silicate layers sandwiching one aluminate layer. These are called 2:1 clays. Other clays such as kaolinite have only one silicate layer to one aluminate layer: these are called 1:1 clays. The various layers are stacked together like cards in a pack to build up the minerals, see Figure 1.

The changes that occur with the removal of more and more silica can be regarded as increasing degrees of weathering. A basalt altered to 2:1 clay minerals is less weathered than one altered to kaolinite, and only extreme weathering can convert a basalt to aluminium hydroxides or bauxite (Figure 2).

Basalt also contains the mineral magnetite, an iron oxide which weathers to iron hydroxide. Iron hydroxides and oxides give the red or yellow colour (rusty colours) to many soils and weathering products.

We have already seen that weathering of a lava-flow is most effective along the cracks or joints. Eventually, rounded chunks of unweathered basalt come to lie in a weathered matrix. These are called corestones, and they frequently have an abrupt, knife-edge contact with weathered rock.

It is also quite common for the corestones to be surrounded by concentric shells of partly weathered rock. This is called spheroidal

WEATHERING OF BASALT

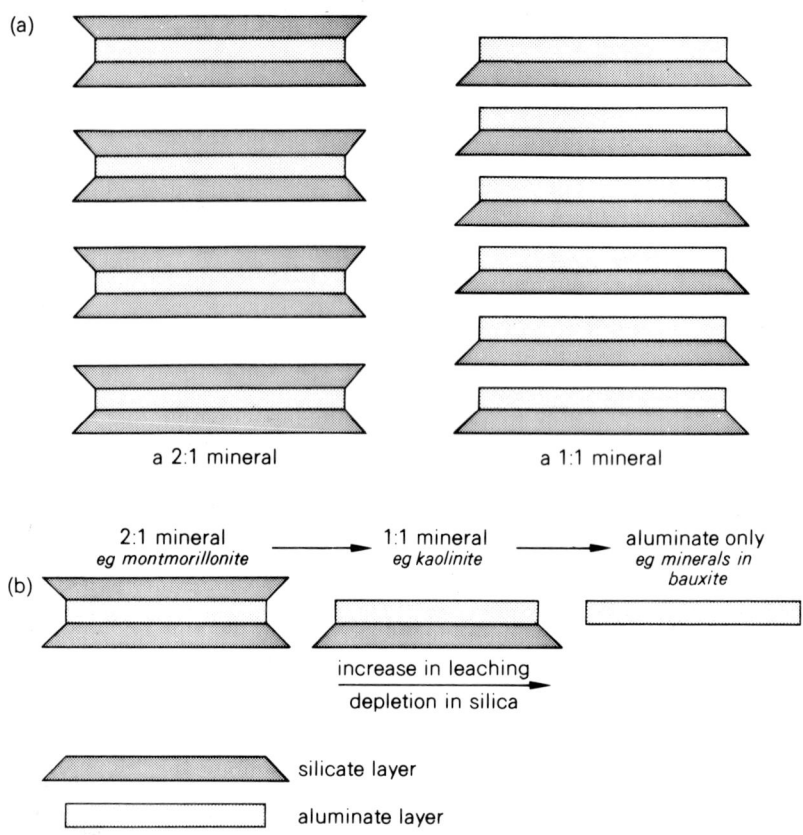

Figure 1 (a) Clay minerals are built of stacks of repeated basic units, which are themselves made of silicate and aluminate layers
(b) Increasing leaching leads to greater loss of silica, so 2:1 minerals give way to 1:1 minerals, which may eventually be replaced by bauxite

weathering. The shells go all around the corestone, so they must be formed where weathering conditions are the same in all directions, probably underground. Rather similar flakes, shells or spalls of rock can be produced on basalt boulders exposed at the ground surface, by processes such as bushfires or hydration, but they do not extend under the boulder. Exfoliation is a general name for the production of layers of weathered rock by any process, including that shown in Figure 3.

A volcano that produces a lava-flow may also build a cone of volcanic ash or cinders. These materials have the same chemical and

WEATHERING OF BASALT

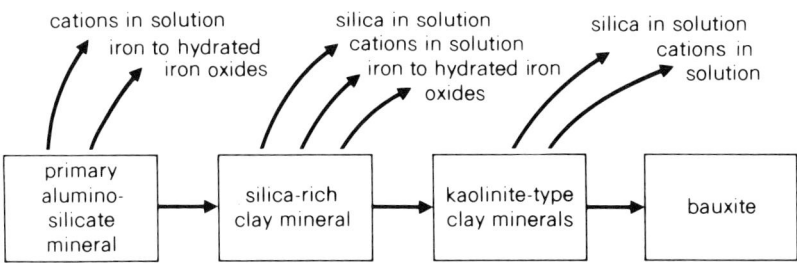

Figure 2 The course of weathering of an alumino-silicate

Figure 3 Spheroidal weathering in basalt, Cooma, New South Wales, Australia. The edges of an original joint block are obvious, and the relationship between the angular corner and the curved weathering features can be seen at the bottom left (photograph by C. D. Ollier)

mineral composition as the lava-flow, but because of the greater ease of penetration by water and air, volcanic ash will usually weather faster than a lava-flow. Indeed basaltic ash can weather in just a few years into a fertile soil.

Some volcanoes, such as the andesite volcanoes that ring the Pacific, emit lavas that are richer in silica than basalt. In general the more silica-rich the lava the slower its weathering will be. The silica-rich lavas may contain quartz (silicon dioxide) which is extremely resistant to weathering.

3. Weathering of limestone

Limestone is largely made up of calcium carbonate in the form of the mineral calcite, and differs from other common rocks in its considerable solubility. Limestone weathers by the solution of calcite, not by its altertion to clay or to other minerals. The landscapes formed on limestones and dominated by solution are so distinctive that they have been given a special name – karst landscapes.

In some limestones the calcite crystals are so tightly packed that the rock is not porous, and solution is confined to the ground surface and to cracks and fissures such as joints and bedding planes between strata.

Surface solution causes irregularities to develop, such as solution grooves (Figure 4) which are commonly called *lapiés* (French) or *karren* (German). If solution is concentrated along joints a series of trenches or grykes may form, with flat platforms or clints in between. Solution concentrated at a point may produce a solution pipe (Figure 5). Large depressions formed in limestone country are known as sink holes or dolines.

Figure 4 Solution grooves on marble (metamorphosed limestone), Takaka Hill, New Zealand (photograph by J. N. Jennings)

WEATHERING OF LIMESTONE

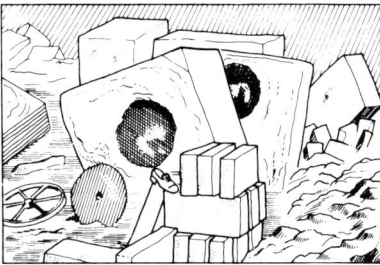

Figure 5 Vertical sections (left) and cross sections (right) of solution pipes exposed in a quarry near Mt Gambier, South Australia, showing how the pipes are formed originally with a clay fill. Presumably, solution proceeds at approximately equal rates in all horizontal directions tending to give circular cross sections, but the most rapid growth is vertically downwards, aided by the movement of water under gravity (sketch by C. D. Ollier)

As fissures in the limestone are widened by solution they become capable of holding more water and eventually all the surface drainage of a limestone area may be replaced by underground drainage (Figure 6). Dry valleys then characterise the surface, and the underground drainage may flow through caves (Figure 7). If the

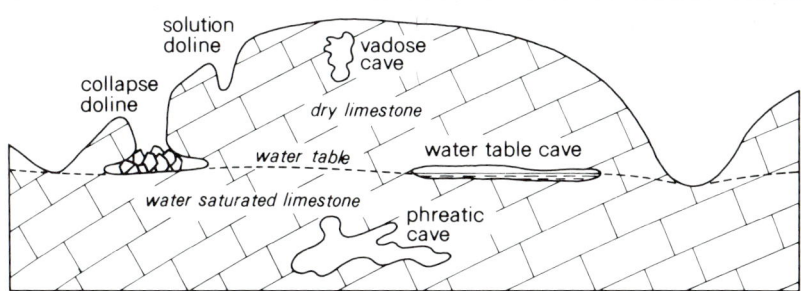

Figure 6 Weathering of limestone in relation to the water table. Pores and cracks in limestone can hold water, and commonly limestones are saturated up to the water table. In the zone above the water table (called the vadose zone), solution by rainwater percolating down may produce features such as solution dolines at the ground surface and vadose caves underground. Some caves are formed in the saturated or phreatic zone; they show evidence of solution in all directions and probably formed very slowly. Underground solution is often most intense just below the water table, where chemical reactions and water flow are both maximised, and caves here are often horizontal, even in steeply dipping rocks. Rockfall into a cave may produce a depression called a collapse doline

WEATHERING OF LIMESTONE

Figure 7 A cave in the coral island Kaileuna, one of the Trobriand Islands, Papua New Guinea. A solutional spongework on the roof suggests that the cave was completely full of water when solution occurred (phreatic conditions). After the cave was drained, precipitation of carbonate built up the stalactites on the roof, the columns on the right and the stalagmite in the centre. The funeral pot and the human bones date back to times when cave burial was practised by the Trobrianders (photograph by D. K. Holdsworth)

roof of a cave collapses, depressions (sink holes) appear at the ground surface).

Impurities in the limestone are not dissolved away but may accumulate as insoluble residues, sometimes producing a distinctive red soil called terra rossa. Large fragments of insoluble rock, such as boulders of sandstone, may protect the underlying limestone from solution and then produce limestone pedestals under perched boulders.

Examples of perched boulders in northern England were emplaced by an ice sheet about 12 000 years ago. The surrounding limestone has been lowered about 50 cm in this time, so the average rate of lowering is about 40 mm in a thousand years.

When the limestone is porous, as in the chalk of England, solution attacks the rock more evenly. But sink holes, caves, and jagged solution forms can be produced in some other forms of porous limestones, such as the limestone of many coral islands, or sand dunes built of fragments of carbonate seashells. It seems that although the process of solution is common to all limestones, details of rock structure control the final landforms.

Some other rocks not generally regarded as soluble can at times be dissolved to produce special landforms. Solution pits (or weathering pits to be more non-committal) are found on granite, quartzite and several other rocks. Grooves similar to solution grooves (karren) on limestone are also found on granite and occasionally other rocks, and are called pseudo-karren (Figure 8).

Figure 8 Solution grooves (pseudo-karren) on granite, Seychelle Islands (photograph by C. D. Ollier)

4. Weathering of granite

Granite differs from basalt in mineral composition and grain size. The mineral grains are several millimetres to several centimetres across and clearly visible to the naked eye. The common minerals are feldspar, mica, and quartz. The feldspars and micas can ultimately weather to clay, but the quartz remains as virtually unweatherable grains. In an advanced state of weathering a granite alters to a mass of clay with dispersed grains of quartz, whereas in the early stages of weathering the boundaries between mineral grains are weathered first, and the rock may crumble even though most of the minerals are little weathered. This is called disintegration, and is widespread in a wide range of environments from the Antartic to hot deserts, but it is uncommon in very wet places where the grains alter completely to clay.

As in basalt, the weathering of granite tends to be concentrated along joints, giving rise to hard corestones embedded in a soft matrix rich in clay. Some varieties of altered granite are given local names, such as growan in south-west England, or grus in the USA. Two technical terms will be found helpful in describing weathering products: saprolite refers to completely altered rock in place; regolith is a less definite term for all the broken or altered weathered material (Figure 9) regardless of whether it is in place or has been transported. As in basalt, the weathering front in granite is commonly very abrupt, and there may be several shells of spheroidal weathering around corestones.

Many landforms in granite areas are produced by a first phase of deep weathering to produce saprolite, followed by a partial stripping of the saprolite. Corestones are then left at the surface and may form boulderfields or tors. Higher zones of weathering generally have rounded corestones (Figure 10) but near the base of weathering angular corestones still fit together and can form tors of locked corestones (Figure 11). It has been suggested that the tors of Dartmoor originate in this way.

Granite is commonly weathered to depths of tens or hundreds of metres, and in deep man-made excavations quite large tors can sometimes be seen with their cover of saprolite intact. It has been suggested that the stripping process can also account for the formation of some inselbergs which may be hundreds of metres high (Figure 12). Several models of landscape formation by stripping of

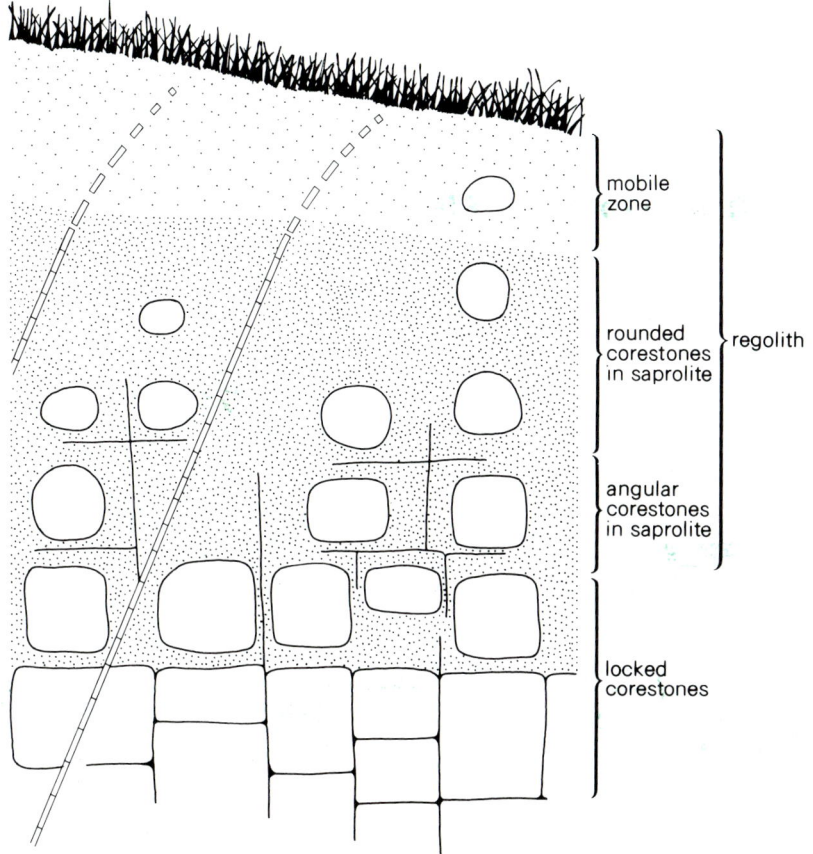

Figure 9 A section through weathered granite showing the relationships of saprolite, regolith and corestones. Quartz veins and some joint planes are preserved undisturbed in the saprolite

regolith have been proposed (Figure 13).

Unlike basalt, granite is formed deep in the earth under conditions of high pressure. When granite is exposed at the ground surface by erosion, the release from pressure causes the rock to expand, a process called unloading. Unloading produces cracks that are roughly parallel to the topography, and most closely spaced near the ground surface, becoming steadily further apart at depth (Figure 14). On convex hills the unloading fractures are convex, in concavities such as glacial cirques they are concave, and beneath plateaus they are roughly horizontal.

WEATHERING OF GRANITE

Figure 10 Road cutting Hong Kong, showing hard unweathered corestones in a soft rotted saprolite in which a joint pattern is still evident (photograph by C. D. Ollier)

Figure 11 A tor. Locked corestones of granite, Devil's Marbles, Northern Territory, Australia (photograph by C. D. Ollier)

WEATHERING OF GRANITE

Figure 12 An inselberg, Lucumbe, near Cubal, Angola. This inselberg rises about 300 m above the surrounding plain (photograph by Ilidio do Amaral)

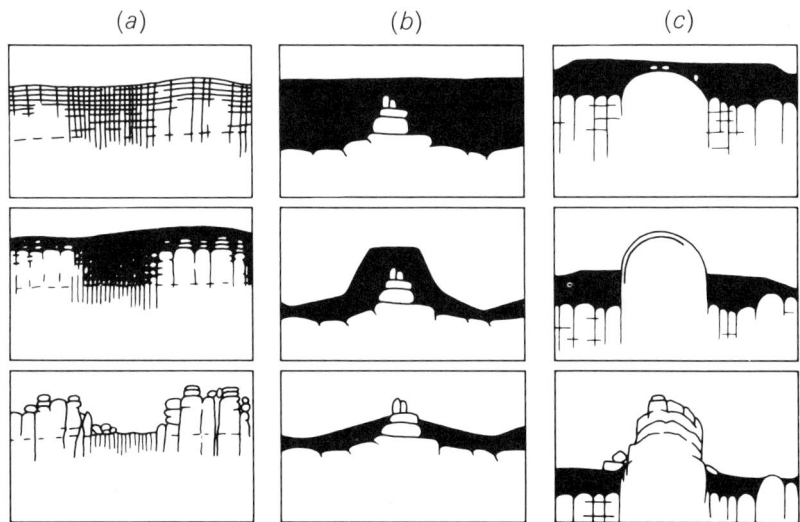

Figure 13 Three models of landscape evolution by stripping of regolith
(a) The formation of tors on Dartmoor according to Linton (1955). The granite is deeply weathered and then the saprolite is completely removed
(b) The formation of inselbergs in Uganda according to Ollier (1960). The granite is deeply weathered and then a new erosion surface is cut across the saprolite, which is not completely removed
(c) The formation of bornhardts (inselbergs) in Nigeria according to Thomas (1965). Each phase of stripping of saprolite is accompanied by further lowering of the weathering front around the bornhardt. In this way the dome may eventually exceed in height the depth of the original weathering profile

WEATHERING OF GRANITE

Figure 14 Sheeting caused by unloading on granitic rock, Sierra Nevada, California (photograph by N. King Huber of the U.S. Geological Survey)

WEATHERING OF GRANITE

The successive sheets produced by unloading are broken up by other weathering processes and dispersed. Rocks can respond to unloading with dangerous suddenness, and some quarries have been forced to close down because of the danger of flying sheets of rock. Many other coarse-grained igneous and metamorphic rocks weather in a similar way to granite, with variation depending on mineral content and rock structure. Unloading can also occur in other rocks which have been subjected to high pressure at some stage in their history. Conglomerate, arkose and limestone all provide good examples.

5. Weathering of sandstone

Sandstones consist of sand grains cemented together by some other material. The weathering of sandstones depends on the weathering of the grains, and also (and usually more importantly) on the weathering of the cement. Most sandstones (arenites) have grains of quartz, a silica mineral that can be regarded as immune to weathering except in extreme conditions. Some sandstones (greywackes) have grains of pre-existing rock or volcanic fragments which are more weatherable; some (calcarenites) have grains of calcium carbonate. Calcarenites weather like limestones; greywackes weather rather like shales.

Quartz sandstones may be cemented by clay, carbonate, iron oxides, silica, or other materials. Those cemented by silica (quartzites or silcretes) are extremely resistant to weathering. Carbonate cement is soluble so carbonate-cemented sandstones weather mainly like limestones. Iron-cemented sandstones are very resistant.

Weathering along joints is common in sandstones. Away from joints sandstone is rather resistant to weathering, and is likely to give rise to joint-bounded hills, perhaps separated by gorges, as in the Bungle Bungles, Western Australia (Figure 15). Despite its resistant appearance, sandstone can also give rise to minor weathering features such as hollows and weather pits.

Figure 15 Spectacular scenery developed on horizontal sandstone in the Bungle Bungle mountains, Western Australia (photograph by C. D. Ollier)

6. Holes, hollows and honeycombs

Weathering can produce a variety of landforms which are basically hollows. Perhaps the simplest is a weathering pit (Figure 16). These are common on granite, but also occur on sandstones and some other rocks. Where some small depression exists, perhaps by some accidental irregularity, it holds water longer than the surroundings and experiences more weathering. A variety of mechanisms may be invoked, including the effects of organisms such as algae that live in the temporary pools. The weathering pits tend to enlarge, getting deeper and wider. They frequently have overflow channels, and several weathering pits may join together. They form best on near-horizontal ground.

Hollows can also form on slopes, sometimes picking out a joint or other area of weakness, but in some instances having no obvious

Figure 16 Weathering pit on granite, Namibia (photograph by C. D. Ollier)

reason to start (Figure 17). On granite these are called tafoni. In some hollows the rock surface is covered by exfoliation scales, but in others the rock is weathering by granular disintegration.

Hollows may grow so big that they can be regarded as caves (Figure 18). It seems that some weathering processes are more effective in the shade than in sunshine, and once a hollow has formed, the shade it induces increases the weathering process. The hollowing of parts of the rock may be associated with case-hardening of the more resistant rock bounding the hollows.

Some case-hardened boulders may be almost entirely hollowed out by weathering (Figure 19). The weathering sometimes attacks the boulders from underneath, and the mechanism is usually granular disintegration. Such features are known in Antarctica, in deserts, and in Mediterranean regions, so there is no direct climatic control. Salt weathering may be associated with at least some of the hollows.

Small hollows are sometimes spaced closely and regularly,

Figure 17 **Weathering hollows (tafoni) on a vertical face of arkose (feldspar-rich sandstone), with no apparent structural cause for their location, Ayers Rock, Central Australia (photograph by C. D. Ollier)**

HOLES, HOLLOWS AND HONEYCOMBS

Figure 18 A cave made by weathering at the base of Ayers Rock, Central Australia. The cave wall is wearing back by flaking, and the cave floor is of bedrock, not sedimentary fill (photograph by C. D. Ollier)

Figure 19 A case-hardened boulder of granite which has been hollowed out by granular disintegration. Kalgoorlie, Western Australia (photograph by C. D. Ollier)

HOLES, HOLLOWS AND HONEYCOMBS

producing honeycomb weathering (Figure 20). Just why the hollows are so regularly arranged is not clear, but honeycomb weathering is particularly common in coastal regions, so salt weathering is possibly involved in some way.

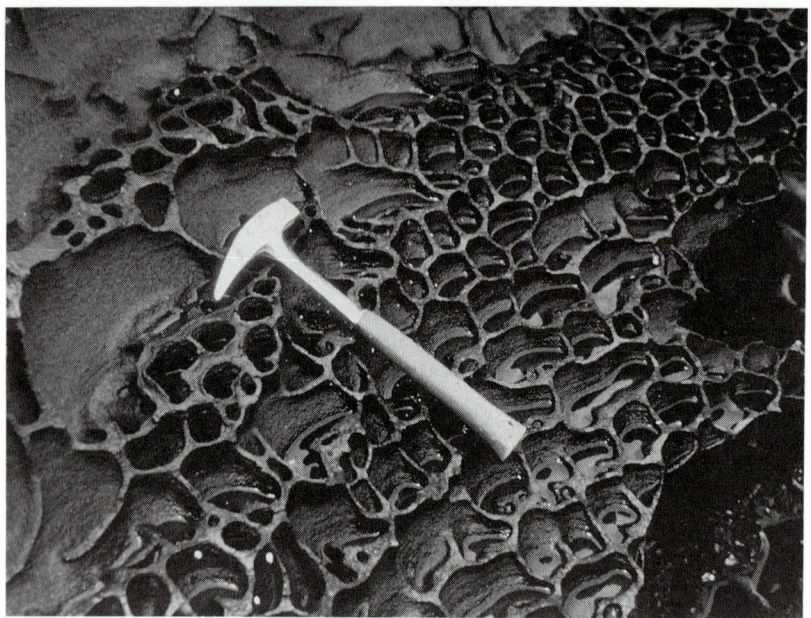

Figure 20 Honeycomb weathering in sandstone at the coast near Durras, New South Wales, Australia (photograph by C. D. Ollier)

7. Weathering at the coast

The sea is salty, and salt weathering is most important around the coast. When salty water dries out the crystallisation of salt prises open cracks and pores, breaking off mineral particles which can be washed away with the next tide. Crystallisation of salt is somewhat like the freezing of ice, in that a solid crystal appears where there was only liquid before, and crystals growing from solution can exert a pressure greater than the cohesion of the enclosing rock. Salt weathering will be most active where conditions of wetting and drying occur most frequently, so its effects decrease rather rapidly away from the shore.

Salt weathering can also occur in arid areas in the vicinity of salt lakes, and can be produced by salts other than common salt, sodium chloride.

Another weathering phenomena important on coasts is wetting-and-drying weathering. Some rocks, especially shales, or other rocks with a significant clay mineral content, simply break up when subjected to alternate wetting and drying out. In tidal regions wetting and drying occur twice a day, giving optimum conditions for this kind of breakdown.

Salt weathering and wetting-and-drying combine to weather rocks that stick up above the water, but both processes are inactive if the rock remains permanently wet. Together they are responsible for the levelling off of any projecting rocks in the tidal zone. If carried far enough this process will produce a shore platform – a flat rock-cut platform that may be hundreds of metres wide but which is covered with a sheet of water no more than a few centimetres deep. Such platforms appear to be commonest in warmer areas: in cold climates erosion by attrition seems to be more important than weathering in coastal geomorphology.

Solution can also be active on coasts, and is particularly evident on limestone coasts (like those of coral islands) where a solution notch may be produced, often overhung by a vizor (Figure 21).

There appears to be a strong tendency for chemical migration of iron in rocks in the coastal zone, and iron hydroxides become concentrated in joints and around concretions. These are then usually more resistant to erosion than the iron-depleted rock between the joints, and differential erosion produces hollows and honeycombs.

Figure 21 A notch and vizor coast formed by solution around sea level on the coral coast of Misima Island, Papua New Guinea (photograph by C. D. Ollier)

8. Weathering in cold climates

The distinctive feature of cold climates, so far as weathering is concerned, is that water is converted into ice. Weathering is largely controlled by the frequency of freeze-thaw alternation. If temperatures are permanently below freezing point there will be little weathering, either chemical or physical, but if temperatures are only sometimes below freezing point weathering will be enhanced.

Figure 22 Frost shattering of a dolerite sill, Born Bluff, Cradle Mountain area, Tasmania. Dolerite is a basic igneous rock with well-developed vertical joints that divide the rock into columns. Frost has made use of these joints and the very coarse debris consists largely of otherwise unweathered joint blocks (photograph by J. A. Peterson)

WEATHERING IN COLD CLIMATES

Water has the peculiar property of expanding on freezing, and ice has about ten per cent more volume than its equivalent in liquid water. It also freezes from the surface down. If water is present in a crack in a rock, its expansion on freezing can widen the crack and eventually lead to fragmentation of the rock (Figure 22).

Mechanical weathering caused by freezing and thawing produces some of the distinctive features of cold climate regions. Jagged and angular frost-shattered rocks are common. Freeze and thaw also assist the flow of debris by solifluction, so that the products of weathering are moved away from weathering sites, exposing fresh surfaces and thus enabling the process to continue. The fragmental products of weathering (Figure 23) accumulate as scree (anywhere), talus (at the foot of a cliff) or 'head', a general term used in England for a layer of debris at the ground surface resulting from former very cold climatic conditions. On more gentle slopes the action of frost tends to sort fragments into stripes or rings of patterned ground (Figure 24).

Figure 23 Frost-weathered upper slopes and scree on lower slopes of a glaciated valley on Mt Baker, Ruwenzori Mountains, Uganda (photograph by C. D. Ollier)

WEATHERING IN COLD CLIMATES

Figure 24 Stone stripes formed of frost-weathered boulders. Rocky Mountains, Colorado (photograph by C. D. Ollier)

Quite large boulders can be prised apart by frost weathering. It has been suggested that the granite tors of Dartmoor have been partly broken apart or 'dismantled' by this process.

Frost can also act on a smaller scale, and could even be responsible for granular disintegration of coarse-grained rocks. But where granulation has been studied, whether in Antarctica, hot deserts, or elsewhere, it is usually found that salt (sodium chloride) is associated with the debris, so perhaps it is salt crystals rather than ice crystals that disrupt the grains.

It is also found in Antartica that the feldspar grains in granites are slightly altered to kaolinite, and that the alteration is taking place at the present time. It would seem that even in generally frozen areas there is sufficient liquid water available for hydrolysis (chemical reaction with water) of feldspars.

9. Weathering in deserts

Deserts are dry areas where water is scarce but seldom totally lacking. Generally we refer to hot deserts, but there are also cold deserts – Antarctica is the driest continent on earth.

In hot deserts one might expect weathering features to be produced by the heat, and especially by the contrast between the hot days and the cold nights that occur in cloudless deserts. Heat causes the expansion of rocks, and also expansion at different rates in different minerals. It seemed quite obvious to early scientists that alternate expansion and contraction caused by daily temperature changes could break down rocks. However, it was discovered in laboratory experiments that repeated heating and cooling produced negligible weathering in dry conditions. Nevertheless there are several landforms, especially split boulders of chemically inert rock, which are most readily explained by heating and cooling, the so-called 'insolation weathering'. Insolation may operate through the widening of cracks by particles of sand or other debris. If a rock expands and the cracks widen, sand particles fall inside: when the rock cools, the crack can no longer close because of the sand wedged in it, so the rock is stressed and the crack extended.

But even in deserts the effects of water are of paramount importance. Water can operate in several ways including moisture-swelling (expansion of rock in a moist atmosphere by mechanisms so far unexplained), crystal growth (in which water is a medium for solution) and most important, hydration. Hydration is the chemical reaction of minerals with water, and in deserts it results in spalls of rusty-coloured flakes of rocks, rather resembling giant corn flakes. In many desert areas almost all rock surfaces are covered by hydration flakes. The spalling of these flakes gradually rounds boulders, and if it occurs on concavities they become deeper and more curved making hollows. This has been called 'negative spheroidal weathering'. In areas where there is no rain it may be that dew is common enough to provide the water required for this process.

Pedestal rocks and balancing rocks are perhaps more common in deserts than in any other climate. The two are really quite different. A balancing rock is not attached to the solid bedrock but is a separate boulder resting, apparently precariously, on an underlying rock. When a rock is perched on top of another, weathering along the contact will proceed in from the edges. The weight of the upper block rests heavily up a point beneath the centre of gravity, and here the two rocks are pressed together closely so that weathering is less effective. There is thus an automatic reduction in weathering beneath the centre of gravity, and the upper rock can end up 'balanced' on a pedestal (Figure 25). Balancing rocks are not confined to arid areas but occur in a wide range of climates (Figure 26).

Pedestal rocks or mushroom rocks, which are projections of solid bedrock, have commonly been attributed to wind erosion by the effect of sand-blasting at the base of a pillar of rock. However, when the rock surfaces are examined in detail it is usually found that there is no trace of sand-blasting, but considerable evidence of such features as flaking or granular disintegration. Such features could not survive sand-blasting, and it seems that most pedestal rocks are produced by weathering (Figure 27), with the wind serving only to remove weathered debris. This is not to deny that sand-blasting can occur in deserts, but its action is confined to a very narrow zone just

Figure 25 Balancing rock, Ungava Gorge River Basin, Labrador. This is granite weathering by granular disintegration in a wind-swept cold area (down to −50°C). The area was deglaciated about 10 000 years ago (photograph by J. A. Peterson)

WEATHERING IN DESERTS

Figure 26 Balanced rock formed in granite, Tenterfield (photograph by New South Wales Information Service)

above the ground surface, and it plays little part in the creation of most desert landforms.

Similarly, many slopes in deserts, as in cliffs, mesas and buttes, have sharp contacts with the flattish floor at their base, and the absence of debris at the base of these cliffs has been taken to indicate inactivity of the landscape. However, it has been shown to be due more probably to rapid removal of debris, and the cliffs may in fact be weathering away quite rapidly. It appears to be significant that in deserts rock debris weathers very rapidly to a particle size that can be removed by wind (or occasional running water).

Chemical weathering in deserts is not so intense as in wetter regions, but we cannot ignore oxidation (giving the red colours of many deserts such as the Red Centre of Australia), hydration, and a certain amount of solution and re-precipitation. Deserts are commonly areas of inland drainage so the weathering products derived locally and those washed in from wetter regions often precipitate.

WEATHERING IN DESERTS

Figure 27 **Pedestal rock. This pillar of Palaeozoic sandstone in the Negev Desert, Israel, was formed by weathering, not wind erosion**

Salts like sodium chloride and carbonates may accumulate in salt lakes and by re-distribution cause salt weathering in neighbouring rocks. Gypsum occurs in many soils, as does calcium carbonate in slightly less arid areas. In places where excess silica is deposited without enough bases to form montmorillonite, secondary silica may be deposited as silcrete. Patches of this may become valuable as precious opal, which has the same composition as common silcrete but has a beautiful colour.

A minor weathering product of deserts is desert varnish, a film of iron and manganese on pebbles that makes them black and shiny. This is not a product of simple chemical weathering, but is produced by algae.

10. Weathering in the humid tropics

The humid tropics have plenty of water and high temperatures, so weathering can be expected to be intense. Indeed mineral alteration is often severe, producing kaolin and iron oxides, and bright red soils and deep weathering profiles are commonly encountered in the tropics. However these weathering products may not result from the present climate alone.

Much of tropical Africa consists of plains with limited erosion and weathering profiles have evolved over a very long time. In many temperate regions on the other hand, land scoured by glaciation during the Ice Age was largely stripped of its weathering mantle. Furthermore, in rugged terrain in the tropics, such as the mountains of New Guinea, while weathering to a few metres is common, very deep weathering is rare: fresh rock is found in active stream courses but is seldom exposed on ridges and slopes. A further complication in relating weathering to present climate is that very deep weathering profiles have now been reported from many places outside the tropics, such as Scotland and Sweden. This weathering is usually attributed to inter-glacial or pre-glacial warm climates. That of Sweden is at least as old as Cretaceous, about 100 million years ago, when climates generally were warm and wet.

Thus some deep weathering profiles have been inherited from long ago and relate to a former climate, not to that of today. Areas that are now temperate or cold had tropical climates in the past. Weathering probably is greater in the humid tropics than elsewhere, but it is difficult to know how much greater because of other factors, such as topography, age of weathering, and duration of weathering.

Different ideas about the effectiveness of tropical weathering lead to different models of landscape evolution. Some suggest that tropical weathering is so intense that tropical rivers have little sand and gravel in their beds, which reduces their ability to erode. Others point out that, at least in the mountainous tropics, rivers have bedloads and erosional ability comparable to that of other climatic regions (Figure 28).

Figure 28 Alluvial gravels in the Strickland River, Papua New Guinea. The large cobbles are about 100 km downstream of their source and clearly show that tropical weathering does not deprive all rivers of gravels to erode their beds. River flows from left to right (photograph by C. D. Ollier)

11. Weathering and life

Plants, animals, bacteria and man himself have a very important part to play in weathering processes and the development of weathering landforms.

Trees widen cracks in rocks by the pressure of their growing roots – they can move heavy boulders, and if they grow near the edge of steep slopes they can cause slope failure. When trees fall over the leverage exerted brings up large masses of soil and rock, exposing fresh rock to surface weathering (Figure 29). Roots are commonly restricted to the top three to six metres, but extreme depths of over fifty metres have been recorded. Larger plants create a distinct microclimate at the ground surface, affecting other forms of weathering.

Roots contribute to chemical weathering. Plants extract nutrients from the soil and return them to the soil in leaf litter, thus cycling many chemical elements. The carbon dioxide content of soil

Figure 29 Columnar-jointed basalt has been prised apart by the growth of tree roots, and when the tree fell the joint blocks were brought up and exposed to atmospheric weathering, Armidale, New South Wales (photograph by J. Field)

WEATHERING AND LIFE

atmospheres is far greater than in the normal atmosphere as a result of respiration of roots, and water that passes through such a soil atmosphere can dissolve more limestone than simple rainwater is able to. In general, vegetation makes soils more acid.

Decaying roots affect the nearby soil and can give rise to pipes made of iron oxides or calcium carbonate by precipitation around the roots. Decaying vegetation, especially leaf litter, also releases organic compounds which are complexing or chelating agents. These enable the movement of various elements, especially iron, through soil profiles and seem to be especially important in podzol soils.

Smaller plants also have weathering effects. Fungi, algae, and lichens (symbiotic associations of algae and fungi) exert a physical effect in enlarging pores, aid in chemical effects by respiration and extraction of nutrients, and by retaining moisture enhance simple chemical weathering. Microfloral assemblages often live in the interior of porous or weathered rock which has been weakened in this way. Surface lichens produce tiny pits. Desert varnish, a shiny film of iron and manganese oxides on pebbles, is produced by the weathering activities of blue-green algae.

Animals may break up rock or mineral fragments by burrowing or by other means, but their main contribution to weathering is the repeated mixing of soil materials, with the consequent exposure of fresh material to weathering agents. Moles, rabbits, prairie dogs and other burrowing animals are 'stirrers', and some make distinct mounds or tunnels.

Termites on the other hand are 'sorters'. They build their termitaria of particles less than a millimetre across, so coarse gravel accumulates at the base of termite activity. This is one possible mechanism for making the 'stone lines' which are common in tropical soils. These are sheets of gravel, seen as lines in cross-sections exposed in trenches or road cuts, which mark the junction between sorted soil above and undisturbed saprolite below.

Earthworms (and some other animals such as millipedes) actually pass earth through their bodies, bringing soil to the surface as worm-casts, and producing a reactive blend of organic and mineral matter. Many other organic weathering mechanisms are known. Snails can wear holes in limestone: bird droppings can corrode limestone and by reacting with it produce new minerals, including economic phosphates. Bat guano reacts similarly, and bats also weather the roofs of caves by mechanical scratching. In the coastal zone, some

WEATHERING AND LIFE

animals bore into rocks, either mechanically or by acid secretions. Respiration by marine vegetation and animals produces carbon dioxide which aids solution of limestone in the coastal zone. Crabs may loosen blocks by jamming themselves into crevices, and some fish actually eat coral.

Man directly affects weathering by ploughing and digging, and indirectly by changing vegetation types either accidentally or by deliberate cultivation. Industrial man produces a lot of waste, and is dependent to a considerable extent on weathering for its decomposition. He also changes atmospheric composition, making it more aggressive. Changes in atmospheric carbon dioxide content may eventually be sufficient to change the world's climate, or the solubility of carbonate in the oceans. Weathering is an important link between man and his environment, and an understanding of this interrelation may contribute to man's ability to survive.

12. Weathering, soil and slopes

Weathering and soil formation are two different but related processes.

Engineers use the term 'soil' to refer to any unconsolidated material, including the entire weathering profile. Soil scientists (pedologists), more interested in agriculture, use 'soil' in a different sense to emphasise 'what plants grow in' and the importance of 'soil-forming processes.' Soil-forming processes change the upper metre or so of the regolith, forming soil horizons which are roughly parallel to the earth's surface. A sequence of soil horizons makes a soil profile and many kinds of soil profiles are recognised by soil scientists.

The soil-forming processes include organic accumulation (such as fallen leaves), organic sorting (as when termites bring fine material to the surface), organic mixing (as by burrowing animals); leaching and precipitation; eluviation (washing out of clay) and illuviation (washing in of clay), and a few other processes. The soil profile may extend to the full depth of the weathering profile, or may only affect the upper part of a much deeper weathering profile. The soil forming processes can be regarded as supplementing normal weathering, or as a specialised aspect of the general weathering process with particular significance near the ground surface.

A few examples of soil types will illustrate the idea of soil profiles.

Podzols

Typical podzol soils occur widely in cold, wet areas, especially under coniferous forest. They are acid soils, have a pale, sandy topsoil overlain by a layer of decaying organic matter and underlain by a horizon enriched in clay and iron oxides (Figures 30 and 31). Possibly clay and iron oxides have been leached and eluviated from the A horizon (aided by chelating agents from decaying pine needles), and have been precipitated and illuviated in the B horizon.

Brown earths

Brown earths occur typically under deciduous forest. They have

WEATHERING, SOIL AND SLOPES

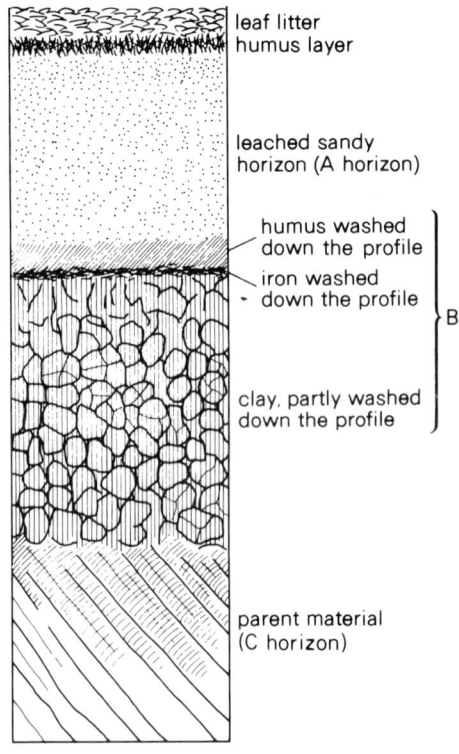

Figure 30 Diagram of a soil profile. This soil is an iron-humus podzol. The leaf litter consists mainly of pine needles, and overlies a layer of slightly decomposed litter, the humus. The clay and iron minerals have been partly redeposited lower in the profile (in the B horizon). In this instance there are layers of accumulation of humus, iron oxides and clay. The soil is formed on a parent material of sedimentary rock which is somewhat weathered towards the top

gradational profiles rather than abrupt horizons, are brown and slightly acid, but there is no movement of iron down the profile. Earthworms repeatedly mix the soil.

Chernozems

Chernozem or black earth soils develop under steppe or prairie grasslands. They have an upper horizon rich in organic matter, which overlies a subsoil in which calcium carbonate accumulates as nodules. The B horizon is a zone of mixing by burrowing animals.

Tropical red earths

The commonest soil in the tropics is a deep red clay, with little variation down the profile, which can simply be called a tropical red

WEATHERING, SOIL AND SLOPES

Figure 31 A typical podzol profile under a forest of birch and conifers, Siberia (photograph by C. D. Ollier)

earth. The soil material is mainly a mixture of kaolin and iron oxides. It is often said that 'laterite' is common in the tropics, but this term is very confused and is best restricted to 'ferricretes' discussed in the section on duricrusts.

The subject of soil science is blighted by the existence of too many soil classifications and soil names. There are some international classifications, and some have been invented for individual countries or by individual soil scientists. The names used above are old-fashioned terms, but have the merit of world-wide understanding.

On slopes weathering may result in leaching of material in solution on the upper slopes and precipitation on lower slopes. This is because upper slopes collect water from rain but lose it to lower sites together with any dissolved chemicals; lower slopes get not only rain, but extra water draining from upslope (with solutes) and so are enriched. This may lead to chemical changes such as precipitation of salt or iron oxides on lower slopes. It is possible for silica and bases to be leached from upper slopes and carried to valley bottoms where they combine to form new clays. This is especially common in tropical areas where upper slopes have red soils rich in iron oxides and kaolin, but the valleys are full of black soils rich in montmorillonite which obviously could not be derived by simple erosion of the valley slopes (Figure 32). The constant repetition of soil units in a landscape, such as red soil on upper slopes and black valley soils is called a catena.

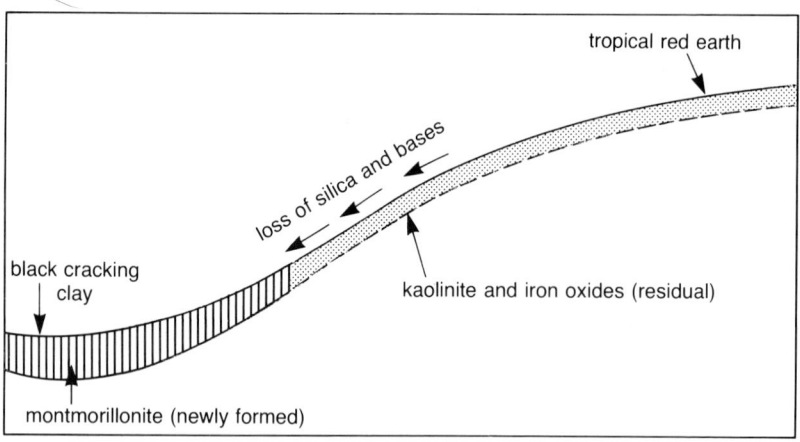

Figure 32 A typical tropical catena. The upper slopes have kaolinite type clays, so the montmorillonite clay in the valleys could not be derived by mechanical erosion from upslope, but results from the creation of new clay minerals in the valley

By an extension of the catena idea we can think of geochemical landscapes. Thus a whole region might be well-watered and leached of chemicals, a neighbouring area may be semi-arid with carbonate precipitation in its soils, and a neighbouring arid area of internal drainage may produce evaporite minerals such as salt from the weathering products it receives from outside its own area.

Duricrusts

Some weathering solutions are transported and precipitated in soils or sediments to form hard layers called duricrusts. They have more

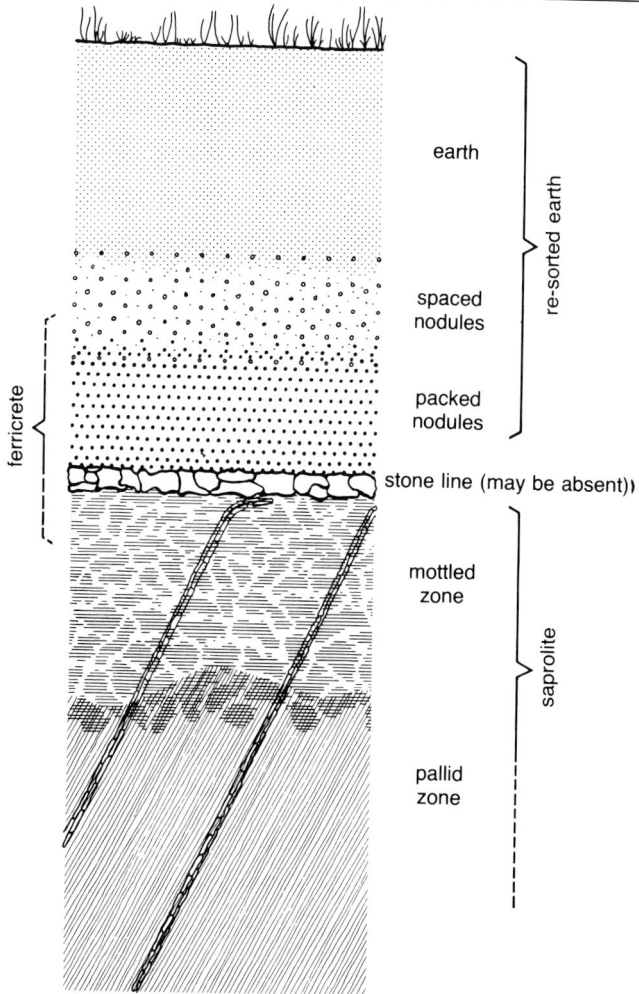

Figure 33 A typical 'laterite' profile, with ferricrete, mottled zone and pallid zone. Saprolite is indicated by quartz veins in place. The laterite used for building in the classic area in India is actually mottled zone material. The ferricrete crust is often formed in re-sorted or transported material

specific names depending on their composition. Ferricrete is rich in iron, calcrete is rich in calcium carbonate, silcrete is rich in silica.

There are many arguments about the origin of silcrete and ferricrete, especially when associated with 'laterite'. Laterite was originally defined as a material which could be dug up and used as bricks. Eventually it came to mean 'ferricrete' to some people, but to others it was equated with tropical red (lateritic) soils. Further investigation showed that 'laterite' is often associated with a weathering profile in which fresh rock at the base is overlain by pale kaolin saprolite (the pallid zone), overlain in turn by a red-and-white mottled zone, and capped by ferricrete (Figure 33). Modern observations suggest that the pallid and mottled zones are typical of many weathering profiles, even those with no ferricrete. The classic laterite material of India is, in fact, the mottled zone of a weathering profile (Figure 34).

Old ideas of ferricrete were concerned with finding a mechanism to bring iron up to the surface where it is obviously concentrated.

Figure 34 A laterite house under construction in Kerala, India, where laterite was first described. The blocks, excavated with an axe-like tool, are stacked for use and some laterite has been left as ready-made walls by careful excavation. On the left a deeper hole dug for a well has brought up pallid zone material (photograph by C. D. Ollier)

WEATHERING, SOIL AND SLOPES

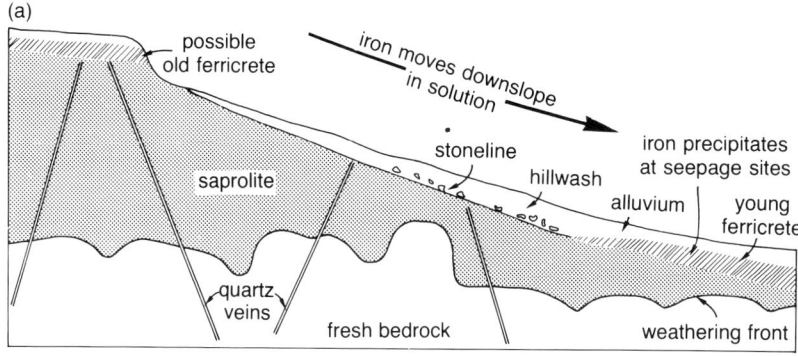

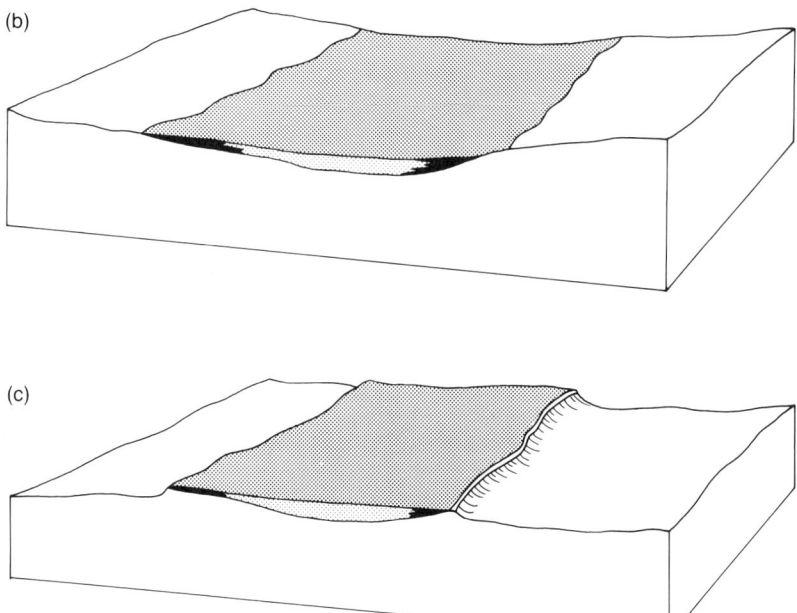

Figure 35 (a) The formation of ferricrete on lower slopes and valley floors by cementation of hillwash and alluvium
(b) In a broad valley with alluvium, iron oxides are precipitated on lower slopes, especially at the edges of alluvium
(c) Erosion of the landscape shown in (b) may result in inversion of relief. Old alluvium is now on hill tops, bounded by ferricrete cliffs or 'breakaways'. Note that the ferricrete need not extend right across the flat hilltop

WEATHERING, SOIL AND SLOPES

Modern ideas suggest that the iron oxide is concentrated on lower slopes or valley bottoms, that is in the lower part of a catena. But ferricrete is commonly found on plateaus. How can it be found on the highest ground if it is formed on lower slopes? The answer lies in inversion of relief, whereby the old ferricreted valley bottoms resist erosion more than the old leached hillsides and eventually are found on hilltops (Figure 35). Many ferricretes, even on hilltops, are formed in sedimentary layers that overlie deep weathered profiles. Similar inversion of relief may affect some calcretes and silcretes.

Duricrusts are of considerable interest in weathering and landscape formation, but at present they are very controversial.

13. Weathering of towns and tombs

The buildings that men erect suffer weathering just as natural rocks do. Sometimes this is good, as when a stone-built house 'mellows' with age and has a more pleasing appearance than when new, but more often the effects of weathering are undesirable.

The actual processes of weathering in towns are similar to those elsewhere, with the addition of increased chemical weathering brought about by polluted atmospheres. Towns, especially industrial towns, have much more carbon dioxide and sulphur dioxide in the atmosphere, which produce acid water that reacts with building stones, especially limestones. It has been said, for example, that the atmosphere of London is slowly and surely converting the Houses of Parliament, built of magnesian limestone, into a 'heap of Epsom Salts'.

Different rocks weather in different ways, and at different rates, and builders are often faced with the dilemma that the best looking stones are the most weatherable, whilst the most durable stone is drab or dull. Small amounts of weatherable but attractive rock may be used to brighten up a building of dull stone, but then there is the danger of 'incompatible weathering' whereby even the normally resistant rock becomes more weatherable where it comes into contact with ornamental stone.

The different weathering qualities of stones may be used structurally in buildings. The toughest stone may be difficult to work and too expensive to use in the whole building. Corners and edges are more exposed to weathering, so should be built of the toughest stone, and cheaper stone used in between. Stone houses in Sheffield, for instance have walls of coal measure sandstone, but the more resistant millstone grit is used on the corners.

The Great Pyramid of Gizeh was built of four kinds of rock. The toughest was used for the edges, two limestones of intermediate quality were used for most of the work, and the most weatherable rock was seldom used at all.

Tombstones reveal the effects of aspect and rock type on weathering and, since they are usually dated, they provide an excellent opportunity to study rates of weathering. It is usually found that limestone and marble tombstones suffer most weathering, sandstone and granite are intermediate, while slate tombstones preserve the details cut on them for the longest period of time. Graveyards provide lots of data for projects in the study of weathering for, besides having stones of different age and rock type, they also provide the opportunity to study the relative weathering of wet and dry sides, sunny and shady sides, contrasts between weathering at the top and base of tombstones, and so on. More elaborate studies may provide quite detailed conclusions, as was the case in an investigation of tombstones covering a 145 year period in Connecticut. It was found that there was an apparent deceleration in weathering during the period 1845–1900 (possibly caused by a general climatic improvement), followed by a substantial acceleration since 1900 amounting to a maximum six-fold increase on north-facing tombstones. Increasing amounts of exhaust gases from industry, vehicles, homes and public buildings are thought to be responsible for the acceleration of weathering (Figure 36).

Figure 36 Weathered tombstones, Middletown, Connecticut, USA, made of arkose (feldspar-rich sandstone). The lower parts of the tombstones are most weathered, because flaking affects the lower, more moist parts of the tombstones most. It is thought that ground moisture is drawn upwards by capillary action (photograph by G. F. Matthias)

WEATHERING OF TOWNS AND TOMBS

On a longer time scale, rates of weathering can be determined from the weathering of antique buildings and statues, and sometimes from suitable geological situations.

The Great Pyramid produced an estimated 50 000 cubic metres of weathered debris in 1 000 years, indicating an average rate of lowering of 0.2 mm per year over the whole surface of the pyramid. At this rate the pyramids will take ten million years to weather away.

14. Weathering products of commercial value

We often think of weathering as a destructive process, but we must remember that it also gives rise to a variety of economic products in which some desirable substance has been concentrated to such an extent that it has commercial value.

In weathering profiles chemical elements migrate and are concentrated in various ways. The upper part of the profile is a zone of leaching and oxidation, and the upper part of this consists of residues which may be of value. The lower part is a reduced and enriched zone where elements derived from above may become concentrated. This is the process of *supergene enrichment*, which was responsible for many of the world's famous mineral deposits.

Clays are the most widespread of residual weathering products. Residual kaolin may be derived by the weathering of a wide range of original rocks: weathered shale can produce nearly pure clay, but weathered granite will contain sand and gravel, for the quartz remains unweathered even when other minerals have been converted entirely to clay. China clay is the purest, whitest and most valuable kaolin used for pottery and for paper filling. It is derived from weathered granite which has to be washed to remove the sand.

The china clay deposits of Cornwall and Devon are the alteration products of granite in place. Their depth is over a hundred metres, which is the economic depth of extraction. It is often maintained that these clays are of hydrothermal origin (made by hot waters coming from below) but since hydrothermal action is a rare process and weathering a common one, and since similar deposits elsewhere in the world are generally thought to be of weathering origin, the hydrothermal hypothesis should not be accepted readily. Modern studies of stable isotopes have confirmed that many very deep clay profiles are formed by weathering.

Several other residual clays have commercial value. The most widely used are the red-burning clays containing iron minerals that are used for making bricks, tiles and pipes. Refractory or fire clays are commonly found under coal seams – they are fossil weathered layers or soils. Bentonite is a weathered volcanic ash consisting largely of montmorillonite, which is put to a wide variety of uses ranging from the mud used in oil drilling to a wine clarifier.

WEATHERING PRODUCTS OF COMMERCIAL VALUE

Residual iron oxide may form on many rocks, including impure limestone, basic igneous rock, serpentine and pyrite-rich rocks, and may be sufficiently concentrated to be of ore quality. Other elements may be concentrated in these residues, as in the nickeliferous laterites of New Caledonia formed on peridotite, or the nickeliferous laterite of Cuba formed on serpentine. Residual manganese is formed in a similar way to residual iron, but in smaller amounts. A particularly useful and distinctive kind of residual material is the 'gossan' formed by weathering of lodes of ore such as lead and copper sulphides. The gossan itself is of no commercial value, but it is used by alert prospectors to locate the valuable minerals beneath.

Bauxite, the ore of aluminium, is a residual deposit formed when practically all other materials, including most silica and iron oxides, have been leached away. Conditions of extreme leaching are necessary for this to happen, and bauxite is generally a weathering product of the humid tropics. Bauxites are especially likely to form over syenites and nepheline syenites – plutonic rocks with very little quartz. They also occur over residual clay derived from limestone, as in Jamaica. Many bauxites are associated with old erosion surfaces, and high relief is not conducive to bauxite formation. Bauxites probably require a long time to form, and many are of early or middle Tertiary age. Some lie on the present ground surface, and the so-called interstratified bauxites were formed on old land surfaces and later buried.

Supergene enrichment is particularly evident in the sulphide ores (Figure 37), especially copper deposits. Many of the world's great copper mines are located on copper-bearing porphyries, where the processes of weathering and supergene enrichment produced bands of copper sulphide of great value. The zone of sulphide enrichment may be only a few metres thick but is commonly tens or hundreds of metres thick and at Bingham, Utah, the deposit is up to 420 m in depth. The oxidised zone likewise may be only a few metres thick, but reaches 600 m at Kennecott, Alaska and 900 m in Zimbabwe. Bearing in mind that these economic deposits are only a particular kind of weathering profile they tell us a great deal about possible depths of weathering. In modern times miners are less interested in the supergene enriched ores than in vast amounts of protore, for, with economies of scale, a very large low grade deposit is preferred to a small high grade deposit. However, in some modern mines such as Bougainville, Papua New Guinea, the concentration of gold in the weathered 'cap' of the deposit has been an economic boon.

WEATHERING PRODUCTS OF COMMERCIAL VALUE

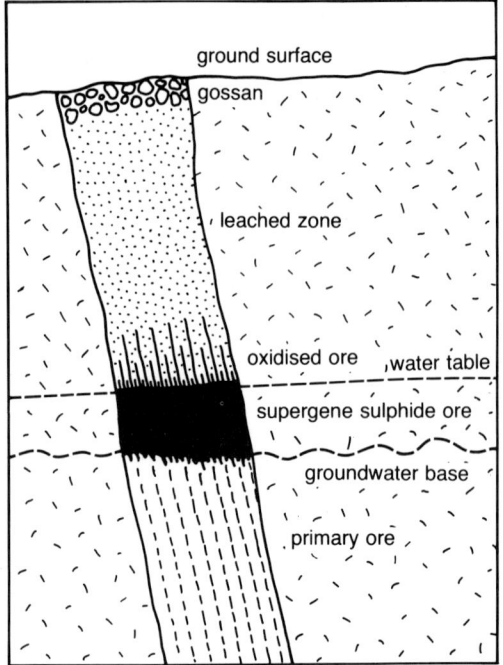

Figure 37 Zones in an ore body formed by supergene enrichment

15. Weathering and landscape evolution

A simplified diagram showing the relationship of weathering to other parts of the geological system is shown in Figure 38. Physical weathering only concerns the upper part of the diagram, with comminution of rock leading to fragmental residue which may then be transported. Chemical weathering results in physical fragmentation and also chemical changes, with losses in solution. The new products of weathering may remain in position as residual deposits. Some, like quartz, may be unaltered. Others are chemically altered

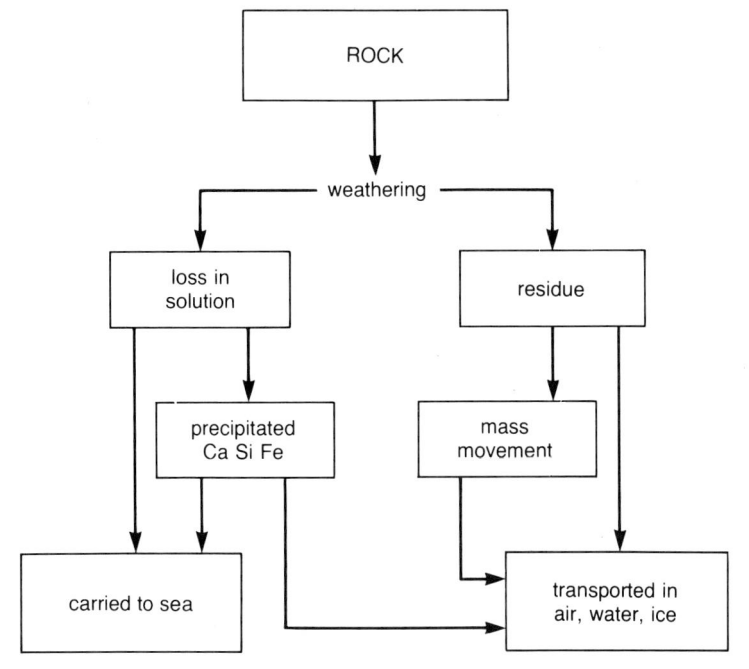

Figure 38 Weathering and weathering products in relation to other landscape-forming processes

WEATHERING AND LANDSCAPE EVOLUTION

but still in place, like residual clay or bauxite. The rest is carried away in solution. Solutions may be carried straight to the sea, or they may be precipitated in soils (perhaps as carbonate nodules, perhaps as duricrusts), in lakes (perhaps as newly formed clays) or in salt lakes (sulphates, chlorides, or other evaporites). The residue will eventually be eroded by rivers, glaciers or mass movements such as creep and landslides. Most weathering is accompanied by erosion and without it there would be few weathering landforms. We talk of weathering landforms where weathering plays a significant part in their genesis, but we must realise that erosion also has an important part to play.

When a river erodes its course through hard rock with little associated weathering, it forms a steep-sided gorge. But most river valleys have much more gentle slopes, and the landscape in areas with rivers is moulded by slope development processes (described in the companion volume *Slope Development*). The slope processes, such as creep and hillwash, operate on broken up, loose material, so

Figure 39 Left and Right Mitten, Monument Valley, Arizona. Different strata weather at different rates, and in conjunction with slope processes affect the form of the landscape as a whole (photograph by C. D. Ollier)

WEATHERING AND LANDSCAPE EVOLUTION

weathering has to prepare the rock for the processes of hillslope evolution as a necessary first stage of landscape evolution (Figure 39).

This vital role of weathering is not confined to regions of simple fluvial geomorphology. In glacial areas frost weathering shatters eroding unweathered rock. In areas of periglacial erosion or in areas dominated by wind transport, weathering plays the same vital role in disintegrating rocks and preparing material suitable for the agents of transportation. In the coastal zone we have seen that the sea can attack unweathered rock, but is more effective where weathering assists in the denudation process.

This brief review of weathering leads to one general conclusion: the most important role of weathering in landscape evolution is not the production of the distinctive though minor landforms described earlier in this book, but the widespread break up and alteration of rock that prepares it for erosion.

Topics for discussion

1. What is the difference between soil and weathered rock? Is there any limit to how thick soils can be?
2. Should the soil be regarded as a lifeless or a dynamic body, or should it be viewed in different ways for different purposes?
3. Discuss the importance of weathering to living things.
4. Is biological weathering anything more than the physical weathering and chemical weathering that happens to be associated with organisms?
5. How do you think rates of weathering will be affected by rates of erosion?
6. Make a list of climates and discuss what kinds of weathering are most likely to prevail in each climatic type.
7. In the geological past conditions for weathering have been very different from those of today. Speculate on the kind of weathering processes that might have occurred (a) before the earth had an oxygen-rich atmosphere, and (b) before plants colonised the land, but after an oxygen-rich atmosphere had evolved.
8. Try to work out what factors are likely to influence the size of caves that may be formed in any given limestone area.
9. Find a world map of bauxite production. Does bauxite appear to be restricted to any particular climate? Under what climatic conditions is bauxite thought to be formed by weathering? Comment on this.
10. Find out what is meant by 'hard water'. What weathering processes may give rise to hard water?

Project work

1. For your local area find out what kind of rocks are present in the high country and in the low country. To what extent is the scenery caused by different rates of weathering of different rocks?

2. Study the tombstones in a local cemetery and see how weathering is related to (a) rock type (b) age (c) orientation (d) height above ground level.

3. Examine the walls of your school and compare the rates of weathering of (a) cement and (b) bricks, stones or other building materials, in different orientation.

4. Examine a coastal area and list all the weathering features you can find. See how they are related to high sea level, and suggest mechanisms for the formation of the weathering forms you describe.

5. Find examples of spheroidal weathering in your local area. Capstones on walls and posts often provide good examples.

6. Examine a deep weathering profile such as may be exposed in a deep excavation or a railway cutting and study the transition from solid rock to soil. Does there appear to be a stage of physical weathering before chemical weathering, or do both processes occur together?

7. Examine soil profiles in your local area on high land and in valleys. Describe the various horizons that make up the soil profiles. Note the colour, feel the texture (e.g. clay, sandy clay, loam, etc.) and note the soil structure (is it crumbly, divided by vertical cracks, broken into rounded lumps, etc.?) Do not be afraid to invent your own descriptive terms at this stage. Try to explain how the soil profile was formed.

8. Shake up a sample of soil in water with a little sodium carbonate (washing soda) to disperse the clay. Allow it to settle and note the different types of sediment and the thickness of the layers. Repeat with samples from different horizons in the same profile and note the changes in the amount of sand down the soil profile.

PROJECT WORK

9 Place a piece of shale, a piece of sandstone and a piece of limestone (or any other three different kinds of rock) into a meshwork container such as a soap-rack. Immerse the specimens in water for a day, then lift out and allow to drain for a day. Repeat this process for several weeks and record the weathering of the different rocks.

10 Place a piece of porous rock in water for a day. Then place it in a freezer for a day. How many times do you need to repeat these operations before the rock starts to distintegrate?

Bibliography and further reading

The only book devoted to weathering with an emphasis on geomorphology is:
Ollier, C. D. 1984 *Weathering*, Longman.

Books that treat weathering with an emphasis on slopes include:
Young, A and Young, D. M. 1990 *Slope Development*, Macmillan.
Small, R. J. and Clark, M. J. 1982 *Weathering and Slopes*, Cambridge University Press.

Weathering and soils are treated in:
Gerrard, A. J. 1981 *Soils and Landforms*, George Allen and Unwin.

Chapters on weathering are found in most textbooks of geomorphology, including:
Derbyshire, E. (ed.) 1976 *Geomorphology and Climate*, Wiley.
Embleton, C. and Thornes, J. (eds.) 1979 *Process in Geomorphology*, Arnold.
Ritter, D. 1978 *Process Geomorphology*, Brown.

Several chapters on applied aspects of weathering studies can be found in:
Cooke, R. U. and Doornkamp, J. C. 1974 *Geomorphology in Environmental Management*, Clarendon Press.
Hails, J. R. (ed.) 1977 *Applied Geomorphology*, Elsevier.

Articles on weathering are scattered through a wide range of journals. Some which have been used in preparing this book are:
Linton, D. L. 1955 'The problem of tors' *Geographical Journal* 121, 470–86
Ollier, C. D. 1960 'The inselbergs of Uganda' *Zeitschrift fur Geomorphologie* 4, 43–52
Thomas, M. F. 1965 'Some aspects of the geomorphology of tors and domes in Nigeria' *Zeitschrift fur Geomorphologie* 9, 63–81